AF095677

THE POETRY OF TERBIUM

The Poetry of Terbium

Walter the Educator

Silent King Books a WhichHead Imprint

Copyright © 2024 by Walter the Educator

All rights reserved. No part of this book may be reproduced in any manner whatsoever without written permission except in the case of brief quotations embodied in critical articles and reviews.

First Printing, 2024

Disclaimer
This book is a literary work; poems are not about specific persons, locations, situations, and/or circumstances unless mentioned in a historical context. This book is for entertainment and informational purposes only. The author and publisher offer this information without warranties expressed or implied. No matter the grounds, neither the author nor the publisher will be accountable for any losses, injuries, or other damages caused by the reader's use of this book. The use of this book acknowledges an understanding and acceptance of this disclaimer.

"Earning a degree in chemistry changed my life!"
– Walter the Educator

dedicated to all the chemistry lovers, like myself, across the world

CONTENTS

Dedication v

Why I Created This Book? 1

One - Oh Terbium 2

Two - Science And Beauty 4

Three - Element Of Enchantment 6

Four - Luminary Of Illumination 8

Five - For All To See 10

Six - World Apart 12

Seven - Knowledge You Proclaim 14

Eight - Celebrate Your Name 16

Nine - Shining So Bright 18

Ten - Forever Impressed 20

Eleven - Terbium, The Luminary 22

Twelve - One Of A Kind 24

Thirteen - Terbium's Might	. . . .	26
Fourteen - Bright And Showered	. . . .	28
Fifteen - Radiant And Rare	. . . .	30
Sixteen - Forever Blessed	. . . .	32
Seventeen - Hidden Power	. . . .	34
Eighteen - Forever Enshrined	. . . . . .	36
Nineteen - Element Of Worth	. . . .	38
Twenty - Relentless Quest	. . . .	40
Twenty-One - Transcends Time	. . . .	42
Twenty-Two - Visionary Force	. . . .	44
Twenty-Three - Towards The Light	. . . . .	46
Twenty-Four - Pushing The Frontier	. . . .	48
Twenty-Five - Continues To Shine	. . . . .	50
Twenty-Six - Never Tire	. . . .	52
Twenty-Seven - Eternal Spring	. . . .	54
Twenty-Eight - Eternally Blessed	. . . . . .	56
Twenty-Nine - Mesmerizing Trance	. . . .	58
Thirty - Cosmic Art	. . . .	60
Thirty-One - Phosphorescent Glory	. . . .	62
Thirty-Two - Brilliance	. . . .	64

Thirty-Three - Wisdom's Impart 66

Thirty-Four - Catalyst For Change 68

Thirty-Five - Stellar Gleam 70

About The Author 72

WHY I CREATED THIS BOOK?

Creating a poetry book about the chemical element Terbium was a unique and fascinating endeavor. Terbium, with its atomic number 65, belongs to the lanthanide series of elements and has several interesting properties. Poetry allows for the exploration and expression of emotions, ideas, and experiences in a creative and imaginative way. By incorporating Terbium as a theme, I can explore its symbolism, scientific properties, historical significance, or even use it as a metaphor for various aspects of life. This book not only combines science and art but also provides a fresh perspective on a lesser-known element, captivating readers with its blend of knowledge and creativity.

ONE

OH TERBIUM

In the depths of the periodic table's realm,
Lies a wondrous element, Terbium by name.
A luminescent beauty, rare and profound,
With secrets and mysteries yet to be found.

Its atomic number, sixty-five it claims,
A symbol Tb, its presence proclaims.
Silvery-white it shines, but don't be deceived,
For within its core, true wonders are conceived.

Terbium, a jewel in the world of science,
With properties that defy all defiance.
Magnetic and luminescent, it holds the key,
To a world of possibilities, waiting to be set free.

In phosphors and lasers, its brilliance shines,
Efficient and vibrant, like celestial signs.
With a green fluorescence, it captivates the eye,
A radiant glow that never wanes, never dies.

Within its atomic structure, electrons dance,
Creating a symphony, a cosmic romance.
In laboratories, scientists delve,
Unlocking the secrets Terbium does tell.

From televisions to computer screens,
Terbium's presence enhances our dreams.
A catalyst of progress, it silently aids,
Advancing technology in countless ways.

Oh Terbium, element of wonder and might,
Your mysteries unfold, shining brightly in sight.
A testament to the beauty of the unknown,
In your atomic embrace, new frontiers are sown.

TWO

SCIENCE AND BEAUTY

In the realm of chemistry's tapestry,
A gem of elements, Terbium, I see.
With atomic number sixty-five it resides,
A rare jewel in nature's hidden tides.

Terbium, a conductor of vibrant light,
In the darkness, its luminescence takes flight.
A phosphorescent spectacle, rare and grand,
Casting an ethereal glow across the land.

Its magnetic allure, a captivating show,
Guiding compass needles where they need to go.
With a magnetic moment, it dances and spins,
Invisible forces at play, where wonder begins.

Terbium, a luminary in the scientific quest,
Revealing the secrets it holds in its chest.

In fluorescent lamps, it lends its vibrant hue,
Illuminating the world, shining bright and true.

    A catalyst for innovation, it plays its part,
Igniting discoveries, fueling the human heart.
In lasers, it finds its purpose and might,
Harnessing energy, piercing through the night.

    Oh Terbium, element of intrigue and awe,
In your atomic dance, new frontiers we draw.
A testament to the wonders of the periodic table,
In your essence, science and beauty intertwine, stable.

THREE

ELEMENT OF ENCHANTMENT

In the realm of elements, Terbium resides,
A mystical force where fascination abides.
With atomic number sixty-five it stands,
A luminescent gem in the alchemist's hands.

Terbium, the conductor of vibrant light,
Invisible hues dance with pure delight.
In the depths of darkness, its glow unfurls,
A celestial symphony that mystically swirls.

With magnetic prowess, it commands the stage,
Guiding compass needles with magnetic sage.
A compass rose, it whispers secrets untold,
Navigating through realms, a story to behold.

Terbium, the luminary of the scientific scene,
Unveiling the secrets hidden within its sheen.

In phosphors and lasers, it finds its role,
Illuminating pathways, igniting the soul.

A catalyst of progress, it fuels innovation's fire,
Igniting discoveries that take us higher.
In the realm of technology, it finds its claim,
Enhancing screens and dreams, igniting a flame.

Oh Terbium, element of enchantment and grace,
Unveiling the mysteries of time and space.
In your atomic dance, a symphony is born,
A testament to the wonders we have yet to adorn.

FOUR

LUMINARY OF ILLUMINATION

In the vast expanse of the periodic table,
A radiant element stands, both rare and stable.
Terbium, its name, resonates with might,
A luminescent star, shining through the night.

Atomic number sixty-five, its claim to fame,
Symbol Tb, it bears, a celestial name.
Silvery-white in appearance, yet it holds,
The secrets of the universe, waiting to unfold.

Terbium, a conductor of light's symphony,
A phosphorescent dance, for all to see.
Within its atomic core, energy thrives,
A vibrant glow, where wonder derives.

Magnetic charisma, it possesses with grace,
Guiding compasses, navigating through space.

An invisible force, a magnetic embrace,
Terbium's influence, a cosmic chase.

From screens to lasers, its purpose is clear,
Illuminating our world, dispelling all fear.
A catalyst for progress, innovation's mate,
Terbium's brilliance, we celebrate.

Oh Terbium, element of luminescent dreams,
In your atomic realm, a symphony gleams.
A testament to the wonders of creation,
Terbium, the luminary of illumination.

FIVE

FOR ALL TO SEE

Terbium, the luminary of the atomic stage,
With atomic number sixty-five, it takes its place.
A rare element, silvery-white and bold,
Its story waiting to be beautifully told.

In the darkness, its luminescence awakes,
A phosphorescent glow that gently shakes.
A celestial dancer, casting vibrant hues,
Terbium's radiance, a mesmerizing muse.

Magnetic in nature, it aligns and guides,
A compass needle's partner, where it resides.
With a magnetic moment, it beckons and leads,
A force unseen, fulfilling navigational needs.

Terbium, catalyst of technological advance,
In labs and industries, it takes a stance.
From fluorescent lamps to computer screens,
Its presence enhances our visual dreams.

An essential player in laser technology,
Terbium's brilliance shines with audacity.
Efficient and powerful, it paves the way,
Pushing boundaries, turning night into day.

Oh Terbium, element of wonder and might,
Your luminescent glow, a captivating sight.
In the realm of science, you hold a key,
Unveiling the mysteries of the universe, for all to see.

SIX

WORLD APART

Terbium, a luminary in the realm of science,
With its atomic magic, it sparks defiance.
A shining star in the periodic table's array,
Unveiling mysteries in its own unique way.

In laboratories, it dances with electrons,
Revealing secrets, like celestial sermons.
Its magnetic charisma, a force to behold,
Guiding compass needles, a story untold.

With fluorescence green, it enchants the night,
A radiant glow, a celestial light.
Phosphors and lasers, its playground of choice,
Igniting innovation with a powerful voice.

As a catalyst, it sets progress ablaze,
Pushing boundaries in countless ways.

In technology's realm, it finds its niche,
Enhancing screens and dreams without a glitch.
   Oh Terbium, element of marvel and allure,
In your atomic dance, dreams find a cure.
A testament to the wonders of science and art,
Terbium, the luminary that sets our world apart.

SEVEN

KNOWLEDGE YOU PROCLAIM

Terbium, the luminary of the atomic stage,
A symphony of electrons, an ethereal sage.
With magnetic allure, it draws the eye,
A captivating dance, that makes us sigh.

In phosphorescent hues, it radiates its light,
Illuminating the dark, a beacon shining bright.
From televisions to smartphones, it finds its place,
Enhancing our visual experience with grace.

A catalyst of innovation, it leads the way,
Unveiling new frontiers, where discoveries sway.
In lasers and sensors, its power is unleashed,
Advancing technology, ensuring progress is reached.

Oh Terbium, element of wonder and might,
In your atomic realm, brilliance takes flight.

A testament to the hidden gems of the periodic table,
Your presence inspires, your essence enables.

From labs to industries, your impact is profound,
Guiding us towards a future where potential is unbound.
Terbium, the luminary, we celebrate your name,
For the gifts you offer, for the knowledge you proclaim.

EIGHT

CELEBRATE YOUR NAME

In the realm of luminescence, a star is born,
Terbium, the element of radiance adorned.
With its phosphorescent glow, an ethereal sight,
Guiding our way through the darkest of night.

In fluorescent lamps, it illuminates the space,
Casting a vibrant glow with elegance and grace.
A catalyst for brilliance, it sparks innovation's flame,
Unleashing new possibilities, igniting a vibrant game.

Within lasers, it dances, a shimmering display,
Harnessing energy, cutting through the fray.
Terbium, the luminary of technology's embrace,
Enhancing screens and dreams, leaving a lasting trace.

Oh Terbium, element of wonder and awe,
In your atomic essence, a symphony we draw.

A testament to the power of science and art,
Guiding us towards a future where brilliance imparts.

With magnetic charisma, it captures our gaze,
Leading us through discoveries, in myriad ways.
Terbium, the luminary, we celebrate your name,
For the light you bring, for the path you proclaim.

NINE

SHINING SO BRIGHT

Terbium, the luminary of the chemical realm,
A dance of electrons, a captivating helm.
Within its atomic core, secrets unfold,
Unveiling wonders, yet to be told.

In phosphorescent hues, it lights up the night,
A cosmic glow, casting shadows in its flight.
A symphony of colors, a celestial ballet,
Terbium's radiance, leading the way.

Magnetic prowess, it holds in its grasp,
Aligning compasses, in a magnetic clasp.
A guide through darkness, a beacon of hope,
Terbium's power, helping us cope.

In lasers it thrives, a source of precision,
Pushing boundaries with scientific vision.
Advancing technology, unlocking doors,
Terbium, the luminary, forever adored.

Oh Terbium, element of brilliance and grace,
Your presence illuminates this cosmic space.
A catalyst for progress, a symbol of might,
Terbium, the luminary, shining so bright.

TEN

FOREVER IMPRESSED

In the realm of elements, a jewel we find,
Terbium, a luminary, one of a kind.
With a phosphorescent glow, it captivates all,
A radiant essence, standing tall.

    Magnetic charisma, it possesses with flair,
Aligning compasses, guiding us with care.
A beacon of direction in the darkest of nights,
Terbium's influence, a celestial light.

    In labs and industries, its purpose unfolds,
Enhancing technologies, breaking molds.
From fluorescent lamps to television screens,
Terbium's brilliance, a vision that gleams.

    A catalyst for progress, with each passing day,
Unveiling wonders in its own unique way.

In lasers and sensors, it paves the path,
Advancing innovation with a powerful aftermath.
    Oh Terbium, element of brilliance and might,
In your atomic dance, dreams take flight.
A testament to science's endless quest,
Terbium, the luminary, we're forever impressed.

ELEVEN

TERBIUM, THE LUMINARY

Terbium, a luminary in the realm of elements,
With its magnetic charm, it beckons and presents.
A vibrant hue, a glow of cosmic delight,
Guiding us through the mysteries of day and night.

In the depths of science, it finds its domain,
A catalyst of discovery, breaking through the mundane.
From laser technology to phosphors aglow,
Terbium's brilliance, a captivating show.

With atomic prowess, it holds the key,
Unlocking secrets of the periodic sea.
In televisions and smartphones, it finds its place,
Enhancing our world with its luminescent grace.

Oh Terbium, element of wonder and awe,
In your atomic dance, science's beauty we adore.

A testament to the marvels of the unseen,
Terbium, the luminary, forever esteemed.
  With each electron spin, a story unfolds,
A symphony of particles, a tale to be told.
Terbium, the lightbearer, shining so bright,
Leading us towards a future bathed in cosmic light.

TWELVE

ONE OF A KIND

In the realm of elements, a gem we find,
Terbium, a luminary, one of a kind.
With atomic charm, it captures our gaze,
A radiant presence that sets us ablaze.

Within its core, a cosmic dance unfurls,
Revealing secrets of the universe, it hurls.
Its fluorescent glow, a celestial hue,
Guiding us through the depths, to something new.

In phosphors and lasers, it finds its place,
A catalyst for progress, with elegant grace.
Enhancing displays, vibrant and clear,
Terbium's touch, the future is near.

Oh Terbium, element of brilliance untamed,
In your atomic symphony, wonders are named.
A testament to science's relentless pursuit,
Terbium, the luminary, forever resolute.

Through innovation's lens, it lights the way,
Unveiling possibilities, day by day.
From labs to industries, its power extends,
Terbium, the luminary, our admiration transcends.

So let us celebrate this element divine,
Terbium, the luminary, forever will shine.
In the realm of science, a beacon of light,
Guiding us towards a future, gleaming and bright.

THIRTEEN

TERBIUM'S MIGHT

Terbium, the luminary of the atomic stage,
A captivating element, with brilliance that won't age.
In the realm of science, it stands tall and proud,
Unveiling mysteries, its secrets it has vowed.

With its magnetic allure, it commands attention,
Aligning compasses, guiding with precision.
A navigator of innovation, it leads the way,
Inventing a future where possibilities sway.

Oh Terbium, element of radiance and might,
In your atomic dance, knowledge takes flight.
A catalyst for progress, a beacon of hope,
Terbium's influence, a gift we can't elope.

Within phosphors, it paints a vivid scene,
Illuminating screens with a vibrant sheen.

From televisions to phones, its light we embrace,
Terbium, the luminary, enlightening our space.

In lasers, it pulses with energy untamed,
Pushing boundaries, where new discoveries are named.
Terbium, the luminary, we celebrate your reign,
For the wonders you've unveiled, for the dreams you sustain.

So let us raise a toast to Terbium's might,
An element of inspiration, forever shining bright.
In the realm of chemistry, a luminary it shall be,
Terbium, the element that sets our spirits free.

FOURTEEN

BRIGHT AND SHOWERED

Terbium, the luminary of the atomic array,
A radiant element, lighting up the way.
In the realm of science, it holds a special place,
Unveiling mysteries with its luminescent grace.

With magnetic allure, it captivates our sight,
Aligning compasses, guiding us day and night.
Terbium, the luminary, a celestial guide,
Leading us through darkness, with magnetism as its stride.

In phosphors and lasers, its purpose is found,
Illuminating screens, with colors so profound.
Terbium, the luminary, a beacon of innovation,
Advancing technologies with vibrant illumination.

Oh Terbium, element of brilliance and might,
In your atomic dance, dreams take flight.

A catalyst for progress, in the scientific quest,
Terbium, the luminary, outshines the rest.
   So let us celebrate this element's allure,
Terbium, the luminary, forever pure.
In the realm of chemistry, a symbol of power,
Guiding us towards a future, bright and showered.

FIFTEEN

RADIANT AND RARE

Terbium, oh radiant and rare,
A luminary beyond compare.
In the realm of chemistry, you grace,
With your atomic elegance, you embrace.

A symphony of electrons, dancing free,
Terbium, the luminary, for all to see.
In fluorescent lights, you cast your glow,
Illuminating spaces, a luminous show.

Magnetic prowess, within your core,
Aligning compasses, forevermore.
A guide through chaos, a steady hand,
Terbium, the luminary, leading us to land.

In lasers, you find your vibrant voice,
Pushing boundaries, giving us choice.
Innovation's ally, bringing forth new day,
Terbium, the luminary, lighting the way.

Oh Terbium, element of wonder and might,
Your brilliance, a source of awe and light.
In the world of science, you're forever adored,
Terbium, the luminary, forever explored.

SIXTEEN

FOREVER BLESSED

In the realm of elements, Terbium gleams,
A luminary in the periodic dreams.
With atomic grace, it weaves its tale,
A story of brilliance, never to fail.

   In phosphors, it paints the night,
With colors vivid, a mesmerizing sight.
Terbium, the luminary, in every hue,
A symphony of light, forever anew.

   In electronic screens, its magic unfolds,
As pixels dance, stories are told.
Terbium, the luminary, a visual delight,
Guiding our senses through the darkest night.

   With magnetic might, it finds its way,
Aligning compasses, nature's ballet.

Terbium, the luminary, a celestial guide,
Leading us towards horizons wide.
    Oh Terbium, element of radiant grace,
In your atomic dance, a masterpiece takes place.
A testament to science's boundless quest,
Terbium, the luminary, forever blessed.

SEVENTEEN

HIDDEN POWER

Terbium, a luminary in our scientific sphere,
With atomic charm and wonders so clear.
In the realm of elements, you hold your sway,
Terbium, the luminary, lighting up the way.

    Your fluorescent prowess, a vibrant display,
In phosphors and screens, you hold sway.
Colors vivid, a mesmerizing dance,
Terbium, the luminary, in radiance enhance.

    Magnetic force, your hidden power,
Guiding compass needles, hour after hour.
Terbium, the luminary, a navigational guide,
Leading us through oceans, far and wide.

    In lasers, you pulse with energy untamed,
Precision and focus, a marvel proclaimed.

Terbium, the luminary, a beacon of light,
Advancing technology with all of your might.
    Oh Terbium, element of brilliance and grace,
In your atomic embrace, we find our place.
A symbol of innovation, a luminary true,
Terbium, the element that shines through.

EIGHTEEN

FOREVER ENSHRINED

Terbium, the luminary, a jewel in the night,
With secrets hidden, in your atomic flight.
A symphony of electrons, dancing in grace,
Terbium, the element, with an enigmatic face.

In phosphor screens, your colors ignite,
Illuminating our world, with a captivating light.
Terbium, the luminary, a painter of dreams,
Weaving vibrant spectrums, like celestial streams.

Magnetic allure, in your magnetic field,
Terbium, the element, an attraction revealed.
Guiding our compass, with a steady hand,
Through uncharted waters, across distant lands.

In lasers, you shine, with a focused beam,
Pushing boundaries, like a visionary dream.
Terbium, the luminary, a pioneer's friend,
Igniting innovation, where possibilities never end.

Oh, Terbium, element of brilliance and might,
In your presence, we find wonder and delight.
A catalyst for progress, a spark for the mind,
Terbium, the luminary, forever enshrined.

NINETEEN

ELEMENT OF WORTH

Terbium, the guardian of luminescent gleam,
In the depths of science, you reign supreme.
With phosphorescent glow, you mesmerize,
A radiant jewel that captures our eyes.

In screens and displays, your colors unfold,
Terbium, the luminary, a story untold.
From TVs to smartphones, you light the way,
Guiding our digital journey, night and day.

Magnetic properties, a force to behold,
Terbium, the luminary, a tale yet untold.
Aligning compasses, pointing true north,
A magnetic dance, a guiding force.

In lasers, you pulsate with vibrant might,
Terbium, the luminary, a celestial light.
Advancing technologies with each beam,
Pushing boundaries, fulfilling our dreams.

Oh Terbium, element of brilliance and grace,
Within your atomic embrace, we find solace.
A symbol of innovation, an element of worth,
Terbium, the luminary, illuminating the Earth.

TWENTY

RELENTLESS QUEST

Terbium, a luminary of the chemical realm,
In your atomic dance, a captivating helm.
Glowing with purpose, in vibrant shades,
A radiant beacon, where brilliance cascades.

    In phosphorescent hues, you come alive,
Painting the world with colors that thrive.
Terbium, the luminary, a visual delight,
Guiding our senses through the darkest night.

    Magnetic allure, within your core,
Aligning compasses, forevermore.
Terbium, the luminary, a guide so true,
Leading us through uncharted avenues.

    In lasers, you sparkle with radiant grace,
A symphony of light, a celestial embrace.

Terbium, the luminary, pushing the frontier,
Advancing technologies, crystal clear.
 Oh Terbium, element of wonder and might,
In your brilliance, a universe takes flight.
A testament to science's relentless quest,
Terbium, the luminary, forever blessed.

TWENTY-ONE

TRANSCENDS TIME

Terbium, the luminary, a jewel in the sky,
Your radiance enchants, catches every eye.
In the palette of elements, you shine so bright,
An iridescent star, a mesmerizing sight.

    Magnetic magic, within your being,
Aligning forces, a symphony of seeing.
Terbium, the luminary, a compass's guide,
Leading us forward, with unyielding stride.

    In the realm of lasers, you find your voice,
A dance of photons, a path of choice.
Terbium, the luminary, a beacon of hope,
Igniting innovation, helping us cope.

    Oh Terbium, element of brilliance and grace,
In your presence, we find solace and embrace.

A symbol of progress, a luminary sublime,
Terbium, the element that transcends time.

TWENTY-TWO

VISIONARY FORCE

Terbium, the luminary, a jewel in disguise,
Glowing with secrets, hidden from prying eyes.
In the depths of science, your story unfolds,
A tale of wonder waiting to be told.

Magnetic master, your power commands,
Aligning the compass in explorers' hands.
Terbium, the luminary, a navigator's aid,
Guiding adventurers on paths yet unswayed.

In phosphor screens, your light comes alive,
Illuminating pixels, where stories thrive.
Terbium, the luminary, a visual symphony,
Enchanting our senses with vivid harmony.

Oh Terbium, element of brilliance and allure,
Within your atomic structure, mysteries endure.

A catalyst for progress, a catalyst for change,
Terbium, the luminary, forever in range.

   In lasers, your beams cut through the night,
Pushing the boundaries of human sight.
Terbium, the luminary, a visionary force,
Fueling innovation, charting a new course.

   So, Terbium, we celebrate your unique role,
An elemental luminary, captivating our soul.
In the realm of science, you shine so bright,
Terbium, the luminary, a beacon of light.

TWENTY-THREE

TOWARDS THE LIGHT

Terbium, the luminary, an element of dreams,
Unveiling a world where imagination gleams.
   Magnetic marvel, your forces align,
Guiding us forward, with purpose defined.
Terbium, the luminary, a compass in the dark,
Leading us on, leaving no room for a spark.
   In lasers, you dance with vibrant grace,
Illuminating paths, in every space.
Terbium, the luminary, a beacon of light,
Igniting innovation, in the darkest of night.
   Oh Terbium, element of brilliance and might,
Within your atomic embrace, we find delight.
A symbol of progress, a luminary true,
Terbium, the element that carries us through.
   In the palette of elements, you stand tall,

A masterpiece of nature, captivating all.
Terbium, the luminary, a gift to behold,
Unveiling the wonders, as stories unfold.
  So Terbium, we honor your unique role,
A luminary element, that touches the soul.
In the world of science, you shine so bright,
Terbium, the luminary, guiding us towards the light.

TWENTY-FOUR

PUSHING THE FRONTIER

Terbium, the luminary, a jewel of the Earth,
A shimmering element of infinite worth.
Within your atomic core, secrets reside,
Unveiling mysteries, with each stride.

Oh Terbium, element of radiance and grace,
In your presence, a vibrant world takes place.
With your fluorescent glow, you captivate,
A symphony of colors, a mesmerizing state.

Magnetic allure, in your magnetic field,
Terbium, the luminary, a power concealed.
Guiding our compass, with unwavering force,
Navigating paths, staying on course.

In lasers, you dazzle with precision and might,
Terbium, the luminary, a celestial light.

Advancing technologies, pushing the frontier,
Opening new horizons, crystal clear.
    Oh Terbium, element of brilliance and insight,
In your essence, innovation takes flight.
A testament to science's relentless quest,
Terbium, the luminary, eternally blessed.

TWENTY-FIVE

CONTINUES TO SHINE

Terbium, the luminary, a jewel in the night,
A symphony of electrons, dancing in pure delight.
With your atomic prowess, you shine so bright,
Guiding our way, a celestial light.

In the palette of elements, you paint vibrant hues,
A rainbow of fluorescence, a spectrum we choose.
Terbium, the luminary, a beacon of grace,
Unveiling the secrets of the atomic space.

Magnetic properties, an enchanting force,
Terbium, the luminary, charting a new course.
Aligning compasses, pointing true north,
A magnetic dance, guiding us forth.

In lasers, you sparkle with a radiant flare,
Terbium, the luminary, a brilliance rare.
Advancing technologies, with each pulsating beam,
Pushing boundaries, fulfilling our dreams.

Oh Terbium, element of brilliance and might,
In your radiance, a symphony takes flight.
A testament to science, a luminary divine,
Terbium, the element that continues to shine.

TWENTY-SIX

NEVER TIRE

Terbium, the luminary, a celestial dance,
In the vast cosmos, you leave us in a trance.
With your fluorescence, colors come alive,
A kaleidoscope of beauty, forever to thrive.
Magnetic allure, within your magnetic core,
Terbium, the luminary, a force to explore.
Guiding our compasses, with unwavering might,
Leading us through darkness, towards the light.
In lasers, you shimmer with ethereal grace,
A symphony of photons, a visual embrace.
Terbium, the luminary, pushing boundaries,
Expanding the frontiers of scientific discoveries.
Oh Terbium, element of brilliance and awe,
In your presence, we find wonder and draw.

A symbol of progress, a luminary divine,
Terbium, the element that continues to shine.

    In the realm of elements, you reign supreme,
A testament to nature's grandest scheme.
Terbium, the luminary, forever we aspire,
To unravel your secrets, and never tire.

TWENTY-SEVEN

ETERNAL SPRING

Terbium, the luminary, a jewel in the sky,
Radiating brilliance as time passes by.
Within your atomic embrace, secrets unfold,
A tapestry of wonders waiting to be told.

In the realm of elements, you stand tall,
A shimmering star, captivating all.
Terbium, the luminary, a beacon of light,
Guiding us forward, with wisdom and might.

In phosphors and screens, your colors ignite,
A vibrant display, a dazzling sight.
Terbium, the luminary, painting the scene,
With hues of green, blue, and serene.

As magnets align, your force takes hold,
Terbium, the luminary, magnetic and bold.
Navigating paths, with precision and grace,
A compass rose, leading us to embrace.

Oh Terbium, element of enchantment and art,
Within your essence, science and beauty impart.
Terbium, the luminary, forever we'll sing,
A symphony of elements, an eternal spring.

TWENTY-EIGHT

ETERNALLY BLESSED

Terbium, the luminary, a radiant celestial star,
With magnetic allure, you guide us from afar.
In the tapestry of elements, you shine so bright,
Terbium, the luminary, a beacon of light.

Your atomic dance, a mesmerizing spectacle,
Pulling us closer, like a magnetic miracle.
In fluorescent hues, you paint the world's canvas,
Terbium, the luminary, a prism of enchantment.

With phosphorescent glow, you mesmerize our eyes,
Revealing the unseen, like a mystical guise.
Terbium, the luminary, unlocking secrets untold,
In the realm of science, your story unfolds.

In lasers, you dazzle with precision and might,
Terbium, the luminary, a celestial light.

Advancing technologies, pushing the frontier,
Opening new horizons, crystal clear.
   Oh Terbium, element of brilliance and insight,
In your essence, innovation takes flight.
A testament to science's relentless quest,
Terbium, the luminary, eternally blessed.

TWENTY-NINE

MESMERIZING TRANCE

Terbium, the luminary of the atomic dance,
A symphony of electrons, a cosmic romance.
In the realm of elements, you hold a special place,
Unveiling your wonders, with elegance and grace.
  With your magnetic might, you pull us in,
Terbium, the luminary, a magnetic kin.
Aligning the compass, guiding our way,
Through uncharted territories, where dreams sway.
  In the world of phosphors, you shine so bright,
Terbium, the luminary, a captivating light.
With your fluorescent glow, a vibrant display,
Illuminating our paths, as we find our own way.
  Oh Terbium, element of brilliance and allure,
A luminary in disguise, so pure.

In lasers, you sparkle, a celestial dance,
Terbium, the luminary, a mesmerizing trance.
　In the palette of elements, you stand tall and grand,
Terbium, the luminary, a gem in nature's hand.
With your unique properties, a treasure to behold,
A beacon of inspiration, forever untold.

THIRTY

COSMIC ART

Terbium, the luminary, a jewel in the night,
Radiating brilliance, a celestial delight.
In the realm of elements, you hold a mystic key,
Unlocking wonders of nature, for all to see.

With your atomic dance, you mesmerize the eye,
Phosphorescent glow, like stars in the sky.
Terbium, the luminary, a cosmic art,
Painting the universe with colors, a work of heart.

In lasers, you dance with precision and grace,
Harnessing energy, illuminating space.
Terbium, the luminary, a conductor of light,
Guiding us towards knowledge, shining so bright.

Oh Terbium, element of magic and might,
In your presence, science takes flight.
Expanding horizons, pushing boundaries anew,
Terbium, the luminary, we look up to you.

In the symphony of elements, your voice stands strong,
A testament to nature's harmonious song.
Terbium, the luminary, forever we'll admire,
Your radiance and beauty, never to tire.

THIRTY-ONE

PHOSPHORESCENT GLORY

Terbium, the luminary, a jewel of the Earth,
A symphony of electrons, a wondrous birth.
In the depths of the periodic table, you reside,
A beacon of brilliance, impossible to hide.

With magnetic charm, you capture our gaze,
Terbium, the luminary, a magnetic maze.
Guiding our compasses, pointing true north,
A magnetic force, showing the way forth.

In phosphorescent glory, you come alive,
Terbium, the luminary, radiant and wise.
With colors that shimmer, like a celestial dance,
You captivate our senses, leaving us in a trance.

Oh Terbium, element of luminescent grace,
In your presence, the world finds its place.

A symbol of innovation, a creator of light,
Terbium, the luminary, shining ever so bright.

In the realm of science, you pave the way,
Terbium, the luminary, leading us astray.
Unveiling the mysteries of the atomic domain,
Terbium, the luminary, forever you shall reign.

THIRTY-TWO

BRILLIANCE

Terbium, the luminary, a star in the night,
A wondrous element, shining so bright.
With atomic number sixty-five, you reside,
In the periodic table, a treasure to confide.

Your electrons dance, in orbital delight,
Terbium, the luminary, a captivating sight.
In phosphors and screens, you radiate,
A vivid display, a spectacle innate.

Oh Terbium, element of rare earth,
In your presence, inspiration gives birth.
A catalyst for innovation and growth,
Terbium, the luminary, we all doth.

In magnets, you align, with magnetic charm,
Terbium, the luminary, keeping us warm.
Navigating the fields, with magnetic force,
Guiding our compasses, staying on course.

Oh Terbium, element of wonder and awe,
In your essence, we find the beauty we saw.
Terbium, the luminary, forever we'll adore,
For you bring brilliance, forevermore.

THIRTY-THREE

WISDOM'S IMPART

Terbium, the luminary, a gem so rare,
Within your atomic structure, a treasure to share.
With a gleam in your eye, a shimmering grace,
You illuminate the world, leaving a luminous trace.

In phosphorescent glory, you come alive,
Terbium, the luminary, radiant and wise.
Your vibrant hues, a kaleidoscope of light,
A symphony of colors, dazzling the sight.

Oh Terbium, element of celestial might,
In your presence, darkness takes flight.
A beacon of knowledge, guiding us through,
Terbium, the luminary, we look up to you.

In the realm of science, you pave the way,
Unveiling the secrets, day by day.
A catalyst for innovation, a catalyst for change,
Terbium, the luminary, forever in range.

Oh Terbium, element of brilliance and art,
In your essence, we find wisdom's impart.
Terbium, the luminary, forever we'll sing,
A tribute to your greatness, an eternal spring.

THIRTY-FOUR

CATALYST FOR CHANGE

Terbium, the luminary, an element divine,
In your atomic symphony, rare and fine.
With electrons spinning, a celestial dance,
You captivate our minds with your magnetic trance.
    Oh Terbium, element of mysteries untold,
In your essence, secrets of the universe unfold.
A conductor of light, you radiate with grace,
Illuminating the cosmos, each vibrant space.
    In phosphorescent glory, you come to life,
Terbium, the luminary, banishing all strife.
With your fluorescent glow, a mesmerizing sight,
You guide us through darkness, like stars in the night.
    Oh Terbium, element of innovation and might,
In labs and factories, you shine so bright.

Advancing technology, pushing the frontier,
Terbium, the luminary, we hold you dear.
    In the palette of elements, you paint a new hue,
Terbium, the luminary, our admiration is true.
A symbol of progress, a catalyst for change,
Terbium, the luminary, forever we'll arrange.

THIRTY-FIVE

STELLAR GLEAM

Terbium, the luminary of the atomic stage,
A gleaming element, captivating with its sage.
In the realm of science, you hold a key,
Unveiling the secrets, for all to see.

With magnetic charm, you dance in our midst,
Aligning the forces, a magnetic twist.
Terbium, the luminary, guiding us true,
Through magnetic fields, we follow you.

In the tapestry of elements, you shine so bright,
A beacon of knowledge, a radiant light.
Terbium, the luminary, forever we'll praise,
For the wisdom you bring, in myriad ways.

In lasers, you dazzle, a celestial display,
Harnessing energy, illuminating our way.
Terbium, the luminary, a symphony of photons,
Creating brilliance, like celestial icons.

Oh Terbium, element of luminescent grace,
In your presence, the world finds its place.
A catalyst for progress, an innovator's dream,
Terbium, the luminary, a stellar gleam.

ABOUT THE AUTHOR

Walter the Educator is one of the pseudonyms for Walter Anderson. Formally educated in Chemistry, Business, and Education, he is an educator, an author, a diverse entrepreneur, and he is the son of a disabled war veteran. "Walter the Educator" shares his time between educating and creating. He holds interests and owns several creative projects that entertain, enlighten, enhance, and educate, hoping to inspire and motivate you.

Follow, find new works, and stay up to date
with Walter the Educator™
at WaltertheEducator.com

www.ingramcontent.com/pod-product-compliance
Lightning Source LLC
LaVergne TN
LVHW052000060526
838201LV00059B/3746